The Memory of Mars

by

Raymond F. Jones

Double9
BOOKS

The Memory of Mars
by Raymond F. Jones

ISBN: 978-93-59323-73-2

Published by

DOUBLE 9 BOOKS

2/13-B, Ansari Road, Daryaganj
New Delhi – 110002
info@double9books.com
www.double9books.com
Tel. 011-40042856

ABOUT THE AUTHOR

Raymond Fisher Jones (15 November 1915 – 24 January 1994) was a science fiction author from the United States. He is most known for his 1952 novel This Island Earth, which was made into the 1955 film of the same name. Jones was born in Salt Lake City, Utah, and has always been a member of the Church of Jesus Christ of Latter-day Saints. In 1994, he died in Sandy, Utah. The majority of Jones' short fiction was published in publications such as Thrilling Wonder Stories, Astounding Stories, and Galaxy during the 1940s, 1950s, and 1960s. Between 1951 and 1978, he authored sixteen novels. His short tale "Rat Race," which initially appeared in the April 1966 edition of Analog Science Fiction and Fact, received a Hugo Award nomination. "Correspondence Course," which first appeared in the April 1945 issue of Astounding Stories, was nominated for a Retro Hugo award for best short story in 1996. Another short story, "The Alien Machine," which first appeared in the June 1949 issue of Thrilling Wonder Stories, was later combined with two other short stories, "The Shroud of Secrecy" and "The Greater Conflict," and expanded into the novel This Island Earth, which inspired the film of the same name. Jones also penned the story that was adapted into a Tales of Tomorrow television episode named "The Children's Room" in 1952.

AREPORTER should be objective even about a hospital. It's his business to stir others' emotions and not let his own be stirred. But that was no good, Mel Hastings told himself. No good at all when it was Alice who was here somewhere, balanced uncertainly between life and death.

Alice had been in Surgery far too long. Something had gone wrong. He was sure of it. He glanced at his watch. It would soon be dawn outside. To Mel Hastings this marked a significant and irrevocable passage of time. If Alice were to emerge safe and whole from the white cavern of Surgery she would have done so now.

Mel sank deeper in the heavy chair, feeling a quietness within himself as if the slow creep of death were touching him also. There was a sudden far distant roar and through the window he saw a streak of brightness in the sky. That would be the tourist ship, the Martian Princess, he remembered.

That was the last thing Alice had said before they took her away from him. "As soon as I'm well again we'll go to Mars for a vacation again, and then you'll remember. It's so beautiful there. We had so much fun—"

Funny, wonderful little Alice—and her strange delusion that she still clung to, that they had taken a Martian vacation in the first year of their marriage. It had started about a year ago, and nothing he could say would shake it. Neither of them had ever been to space.

He wished now he had taken her. It would have been worth it, no matter what its personal cost. He had never told her about the phobia that had plagued him all his life, the fear of outer space that made him break out in a cold sweat just to think of it—nor of the nightmare that came again and again, ever since he was a little boy.

There must have been some way to lick this thing—to give her that vacation on Mars that she had wanted so much.

Now it was too late. He knew it was too late.

THE white doors opened, and Dr. Winters emerged slowly. He looked at Mel Hastings a long time as if trying to remember who the reporter was. "I must see you—in my office," he said finally.

Mel stared back in numb recognition. "She's dead," he said.

Dr. Winters nodded slowly as if in surprise and wonder that Mel had divined this fact. "I must see you in my office," he repeated.

Mel watched his retreating figure. There seemed no point in following. Dr. Winters had said all that need be said. Far down the corridor the Doctor turned and stood patiently as if understanding why Mel had not followed, but determined to wait until he did. The reporter stirred and rose from the chair, his legs withering beneath him. The figure of Dr. Winters grew larger as he approached. The morning clatter of the hospital seemed an ear-torturing shrillness. The door of the office closed and shut it out.

"She is dead." Dr. Winters sat behind the desk and folded and unfolded his hands. He did not look at Mel. "We did everything we could, Mr. Hastings. Her injuries from the accident were comparatively minor—" He hesitated, then went on. "In normal circumstances there would have been no question—her injuries could have been repaired."

"What do you mean, 'In normal circumstances—'?"

Dr. Winters turned his face away from Mel for a moment as if to avoid some pain beyond endurance. He passed a weary hand across his forehead and eyes and held it there a moment before speaking. Then he faced Mel again. "The woman you brought in here last night—your wife—is completely un-normal in her internal structure. Her internal organs cannot even be identified. She is like a being of some other species. She is not—she is simply not human, Mr. Hastings."

Mel stared at him, trying to grasp the meaning of the words. Meaning would not come. He uttered a short, hysterical laugh that was like a bark. "You're crazy, Doc. You've completely flipped your lid!"

Dr. Winters nodded. "For hours during the night I was in agreement with that opinion. When I first observed your wife's condition I was convinced I was utterly insane. I called in six other men to verify my observation. All of them were as stupefied as I by what we saw. Organs that had no place in a human structure. Evidence of a chemistry that existed in no living being we had ever seen before—"

The Doctor's words rolled over him like a roaring surf, burying, smothering, destroying—

"I want to see." Mel's voice was like a hollow cough from far away. "I think you're crazy. I think you're hiding some mistake you made yourself. You killed Alice in a simple little operation, and now you're trying to get out of it with some crazy story that nobody on earth would ever believe!"

"I want you to see," said Dr. Winters, rising slowly. "That's why I called you in here, Mr. Hastings."

MEL trailed him down the long corridor again. No words were spoken between them. Mel felt as if nothing were real anymore.

They went through the white doors of Surgery and through the inner doors. Then they entered a white, silent—cold—room beyond.

In the glare of icy white lights a single sheeted figure rested on a table. Mel suddenly didn't want to see. But Dr. Winters was drawing back the cover. He exposed the face, the beloved features of Alice Hastings. Mel cried out her name and moved toward the table. There was nothing in her face to suggest she was not simply sleeping, her hair disarrayed, her face composed and relaxed as he had seen her hundreds of times.

"Can you stand to witness this?" asked Dr. Winters anxiously. "Shall I get you a sedative?"

Mel shook his head numbly. "No—show me ..."

The great, fresh wound extended diagonally across the abdomen and branched up beneath the heart. The Doctor grasped a pair of small scissors and swiftly clipped the temporary sutures. With forceps and retractors he spread open the massive incision.

Mel closed his eyes against the sickness that seized him.

"Gangrene!" he said. "She's full of gangrene!"

Below the skin, the surface layers of fatty tissue, the substance of the tissue changed from the dark red of the wounded tissue to a dark and greenish hue that spoke of deadly decay.

But Dr. Winters was shaking his head. "No. It's not gangrene. That's the way we found the tissue. That appears to be its—normal condition, if you will."

Mel stared without believing, without comprehending.

Dr. Winters probed the wound open further. "We should see the stomach here," he said. "What is here where the stomach should be I cannot tell you. There is no name for this organ. The intestinal tract should lie here. Instead, there is only this homogeneous mass of greenish, gelatinous material. Other organs, hardly differentiated from this mass, appear where the liver, the pancreas, the spleen should be."

Mel was hearing his voice as if from some far distance or in a dream.

"There are lungs—of a sort," the Doctor went on. "She was certainly capable of breathing. And there's a greatly modified circulatory system, two of them, it appears. One circulates a blood substance in the outer layers of tissue that is almost normal. The other circulates a liquid that gives the remainder of the organs their greenish hue. But how circulation takes place we do not know. She has no heart."

MEL HASTINGS burst out in hysterical laughter. "Now I know you're crazy Doc! My tender, loving Alice with no heart! She used to tell me, 'I haven't got any brains. I wouldn't have married a dumb reporter if I did. But so I've got a heart and that's what fell in love with you—my heart, not my brains.' She loved me, can't you understand that?"

Dr. Winters was slowly drawing him away. "I understand. Of course I understand. Come with me now, Mr. Hastings, and lie down for a little while. I'll get you something to help take away the shock."

Mel permitted himself to be led away to a small room nearby. He drank the liquid the Doctor brought, but he refused to lie down.

"You've shown me," he said with dull finality. "But I don't care what the explanation is. I knew Alice. She was human all right, more so than either you or I. She was completely normal, I tell you—all except for this idea she had the last year or so that we'd gone together on a vacation to Mars at one time."

"That wasn't true?"

"No. Neither of us had ever been out in space."

"How well did you know your wife before you married her?"

Mel smiled in faint reminiscence. "We grew up together, went to the same grade school and high school. It seems like there was never a time when Alice and I didn't know each other. Our folks lived next door for years."

"Was she a member of a large family?"

"She had an older brother and sister and two younger sisters."

"What were her parents like?"

"They're still living. Her father runs an implement store. It's a farm community where they live. Wonderful people. Alice was just like them."

Dr. Winters was silent before he went on. "I have subjected you to this mental torture for just one reason, Mr. Hastings. If it has been a matter of any less importance I would not have told you the details of your wife's condition, much less asking you to look at her. But this is such an enormous scientific mystery that I must ask your cooperation in helping to solve it. I want your permission to preserve and dissect the body of your wife for the cause of science."

Mel looked at the Doctor in sudden sharp antagonism. "Not even give her a burial? Let her be put away in bottles, like—like a—"

"Please don't upset yourself any more than necessary. But I do beg that you consider what I've just proposed. Surely a moment's reflection will show you that this is no more barbaric than our other customs regarding our dead.

"But even this is beside the point. The girl, Alice, whom you married is like a normal human being in every apparent external respect, yet the organs which gave her life and enabled her to function are like nothing encountered before in human experience. It is imperative that we understand the meaning of this. It is yours to say whether or not we shall have this opportunity."

Mel started to speak again, but the words wouldn't come out.

"Time is critical," said Dr. Winters, "but I don't want to force you to an instantaneous answer. Take thirty minutes to think about it. Within that time, additional means of preservation must be taken. I regret that I must be in such haste, but I urge that your answer be yes."

Dr. Winters moved towards the door, but Mel gestured for him to remain.

"I want to see her again," Mel said.

"There is no need. You have been tortured enough. Remember your wife as you have known her all her life, not as you saw her a moment ago."

"If you want my answer let me see her again."

Dr. Winters led the way silently back to the cold room. Mel drew down the cover only far enough to expose the face of Alice. There was no mistake. Somehow he had been hoping that all this would turn out to be some monstrous error. But there was no error.

Would she want me to do what the Doctor has asked? he thought. She wouldn't care. She would probably think it a very huge joke that she had been born with innards that made her different from everybody else. She would be amused by the profound probings and mutterings of the learned doctors trying to find an explanation for something that had no explanation.

Mel drew the sheet tenderly over her face.

"You can do as you wish," he said to Dr. Winters. "It makes no difference to us — to either of us."

THE sedative Dr. Winters had given him, plus his own exhaustion, drove Mel to sleep for a few hours during the afternoon, but by evening he was awake again and knew that a night of sleeplessness lay ahead of him. He couldn't stand to spend it in the house, with all its fresh reminders of Alice.

He walked out into the street as it began to get dark. Walking was easy; almost no one did it any more. The rush of private and commercial cars swarmed overhead and rumbled in the ground beneath. He was an isolated anachronism walking silently at the edge of the great city.

He was sick of it. He would have liked to have turned his back on the city and left it forever. Alice had felt the same. But there was nowhere to go. News reporting was the only thing he knew, and news occurred only in the great, ugly cities of the world. The farmlands, such as he and Alice had known when they were young, produced nothing of interest to the satiated denizens of the towns and cities. Nothing except food, and much of this was now being produced by great factories that synthesized protein and carbohydrates. When fats could be synthesized the day of the farmer would be over.

He wondered if there weren't some way out of it now. With Alice gone there was only himself, and his needs were few. He didn't know, but suddenly he wanted very much to see it all again. And, besides, he had to tell her folks.

THE ancient surface bus reached Central Valley at noon the next day. It all looked very much as it had the last time Mel had seen it and it looked very good indeed. The vast, open lands; the immense ripe fields.

The bus passed the high school where Mel and Alice had attended classes together. He half expected to see her running across the campus lawn to meet him. In the middle of town he got off the bus and there were Alice's mother and father.

They were dry-eyed now but white and numb with shock. George Dalby took his hand and pumped it heavily. "We can't realize it, Mel. We just can't believe Alice is gone."

His wife put her arms around Mel and struggled with her tears again. "You didn't say anything about the funeral. When will it be?"

Mel swallowed hard, fighting the one lie he had to tell. He almost wondered now why he had agreed to Dr. Winters' request. "Alice—always wanted to do all the good she could in the world," he said. "She figured that she could be of some use even after she was gone. So she made an agreement with the research hospital that they could have her body after she died."

It took a moment for her mother to grasp the meaning. Then she cried out, "We can't even bury her?"

"We should have a memorial service, right here at home where all her friends are," said Mel.

George Dalby nodded in his grief. "That was just like Alice," he said. "Always wanting to do something for somebody else—"

And it was true, Mel thought. If Alice had supposed she was not going to live any longer she would probably have thought of the idea, herself. Her parents were easily reconciled.

They took him out to the old familiar house and gave him the room where he and Alice had spent the first days of their marriage.

WHEN it was night and the lights were out he felt able to sleep naturally for the first time since Alice's accident. She seemed not far away here in this old familiar house.

In memory, she was not, for Mel was convinced he could remember the details of his every association with her. He first became conscious of her existence one day when they were in the third grade. At the beginning of each school year the younger pupils went through a course of weighing, inspection, knee tapping, and cavity counting. Mel had come in late for his examination that year and barged into the wrong room. A shower of little-girl squeals had greeted him as the teacher told him kindly where the boy's examination room was.

But he remembered most vividly Alice Dalby standing in the middle of the room, her blouse off but held protectingly in front of her as she jumped up and down in rage and pointed a finger at him. "You get out of here, Melvin Hastings! You're not a nice boy at all!"

Face red, he had hastily retreated as the teacher assured Alice and the rest of the girls that he had made a simple mistake. But how angry Alice had been! It was a week before she would speak to him.

He smiled and sank back deeply into the pillow. He remembered how proud he had been when old Doc Collins, who came out to do the honors every Fall, had told him there wasn't a thing wrong with him and that if he continued to drink his milk regularly he'd grow up to be a football player. He could still hear Doc's words whistling through his teeth and feel the coldness of the stethoscope on his chest.

Suddenly, he sat upright in bed in the darkness.

Stethoscope!

They had tapped and inspected and listened to Alice that day, and all the other examination days.

If Doc Collins had been unable to find a heartbeat in her he'd have fainted — and spread the news all over town!

Mel got up and stood at the window, his heart pounding. Old Doc Collins was gone, but the medical records of those school examinations might still be around somewhere. He didn't know what he expected to prove, but surely those records would not tell the same story Dr. Winters had told.

It took him nearly all the next day. The grade school principal agreed to help him check through the dusty attic of the school, where ancient records and papers were tumbled about and burst from their cardboard boxes.

Then Paul Ames, the school board secretary, took Mel down to the District Office and offered to help look for the records. The old building was stifling hot and dusty with summer disuse. But down in the cool, cobwebbed basement they found it.... Alice's records from the third grade on up through the ninth. On every one: heart, o.k.; lungs, normal. Pulse and blood pressure readings were on each chart.

"I'd like to take these," said Mel. "Her doctor in town—he wants to write some kind of paper on her case and would like all the past medical history he can get."

Paul Ames frowned thoughtfully. "I'm not allowed to give District property away. But they should have been thrown out a long time ago—take 'em and don't tell anybody I let you have 'em."

"Thanks. Thanks a lot," Mel said.

And when she was fourteen or fifteen her appendix had been removed. A Dr. Brown had performed the operation, Mel remembered. He had taken over from Collins.

"Sure, he's still here," Paul Ames said. "Same office old Doc Collins used. You'll probably find him there right now."

Dr. Brown remembered. He didn't remember the details of the appendectomy, but he still had records that showed a completely normal operation.

"I wonder if I could get a copy of that record and have you sign it," Mel said. He explained about the interest of Dr. Winters in her case without revealing the actual circumstances.

"Glad to," said Dr. Brown. "I just wish things hadn't turned out the way they have. One of the loveliest girls that ever grew up here, Alice."

The special memorial service was held in the old community church on Sunday afternoon. It was like the drawing of a curtain across a portion of Mel's life, and he knew that curtain would never open again.

He took a bus leaving town soon after the service.

There was one final bit of evidence, and he wondered all the way back to town why he had not thought of it first. Alice's pregnancy had ended in miscarriage, and there had never been another.

But X-rays had been taken to try to find the cause of Alice's difficulty. If they showed that Alice was normal within the past two years—

DR. WINTERS was mildly surprised to see Mel again. He invited the reporter in to his office and offered him a chair. "I suppose you have come to inquire about our findings regarding your wife."

"Yes — if you've found anything," said Mel. "I've got a couple of things to show you."

"We've found little more than we knew the night of her death. We have completed the dissection of the body. A minute analysis of each organ is now under way, and chemical tests of the body's substances are being made. We found that differences in the skeletal structure were almost as great as those in the fleshy tissues. We find no relationship between these structures and those of any other species — human or animal — that we have ever found."

"And yet Alice was not always like that," said Mel.

Dr. Winters looked at him sharply. "How do you know that?"

Mel extended the medical records he had obtained in Central Valley. Dr. Winters picked them up and examined them for a long time while Mel watched silently.

Finally, Dr. Winters put the records down with a sigh. "This seems to make the problem even more complex than it was."

"There are X-rays, too," said Mel. "Alice had pelvic X-rays only a little over two years ago. I tried to get them, but the doctor said you'd have to request them. They should be absolute proof that Alice was different then."

"Tell me who has them and I'll send for them at once."

An hour later Dr. Winters shook his head in disbelief as he turned off the light box and removed the X-ray photograph. "It's impossible to believe that these were taken of your wife, but they corroborate the evidence of the other medical records. They show a perfectly normal structure."

The two men remained silent across the desk, each reluctant to express his confused thoughts. Dr. Winters finally broke the silence. "It must be, Mr. Hastings," he said, " — it must be that this woman — this utterly alien person — is simply not your wife, Alice. Somehow, somewhere, there must be a mistake in identity, a substitution of similar individuals."

"She was not out of my sight," said Mel. "Everything was completely normal when I came home that night. Nothing was out of place. We went out to a show. Then, on the way home, the accident occurred. There could have been no substitution — except right here in the hospital. But I know it was Alice I saw. That's why I made you let me see her again — to make sure."

"But the evidence you have brought me proves otherwise. These medical records, these X-rays prove that the girl, Alice, whom you married, was quite normal. It is utterly impossible that she could have metamorphosed into the person on whom we operated."

Mel stared at the reflection of the sky in the polished desk top. "I don't know the answer," he said. "It must not be Alice. But if that's the case, where is Alice?"

"That might even be a matter for the police," said Dr. Winters. "There are many things yet to be learned about this mystery."

"There's one thing more," said Mel. "Fingerprints. When we first came here Alice got a job where she had to have her fingerprints taken."

"Excellent!" Dr. Winters exclaimed. "That should give us our final proof!"

It took the rest of the afternoon to get the fingerprint record and make a comparison. Dr. Winters called Mel at home to give him the report. There was no question. The fingerprints were identical. The corpse was that of Alice Hastings.

THE nightmare came again that night. Worse than Mel could ever remember it. As always, it was a dream of space, black empty space, and he was floating alone in the immense depths of it. There was no direction. He was caught in a whirlpool of vertigo from which he reached out with agonized yearning for some solidarity to cling to.

There was only space.

After a time he was no longer alone. He could not see them, but he knew they were out there. The searchers. He did not know why he had to flee or why they sought him, but he knew they must never overtake him, or all would be lost.

Somehow he found a way to propel himself through empty space. The searchers were growing points of light in the far distance. They gave him a sense of direction. His being, his existence, his universe of meaning and understanding depended on the success of his flight from the searchers. Faster, through the wild black depths of space—

He never knew whether he escaped or not. Always he awoke in a tangle of bedclothes, bathed in sweat, whimpering in fear. For a long time Alice had been there to touch his hand when he awoke. But Alice was gone now and he was so weary of the night pursuit. Sometimes he wished it would end with the searchers—whoever they were—catching up with him and doing what they intended to do. Then maybe there would be no more nightmare. Maybe there would be no more Mel Hastings, he thought. And that wouldn't be so bad, either.

He tossed sleeplessly the rest of the night and got up at dawn feeling as if he had not been to bed at all. He would take one day more, and then get back to the News Bureau. He'd take this day to do what couldn't be put off any longer — the collecting and disposition of Alice's personal belongings.

HE shaved, bathed and dressed, then began emptying the drawers, one by one. There were many souvenirs, mementos. She was always collecting these. Her bottom drawer was full of stuff that he'd glimpsed only occasionally.

In the second layer of junk in the drawer he came across the brochure on Martian vacations. It must have been one of the dreams of her life, he thought. She'd wanted it so much that she'd almost come to believe that it was real. He turned the pages of the smooth, glossy brochure. Its cover bore the picture of the great Martian Princess and the blazoned emblem of Connemorra Space Lines. Inside were glistening photos of the plush interior of the great vacation liner, and pictures of the domed cities of Mars where Earthmen played more than they worked. Mars had become the great resort center of Earth.

Mel closed the book and glanced again at the Connemorra name. Only one man had ever amassed the resources necessary to operate a private space line. Jim Connemorra had done it; no one knew quite how. But he operated now out of both hemispheres with a space line that ignored freight and dealt only in passenger business. He made money, on a scale that no government-operated line had yet been able to approach.

Mel sank down to the floor, continuing to shift through the other things in the drawer.

His hand stopped. He remained motionless as recognition showered sudden frantic questions in his mind. There lay a ticket envelope marked Connemorra Lines.

The envelope was empty when he looked inside, and there was no name on it. But it was worn. As if it might have been carried to Mars and back.

In sudden frenzy he began examining each article and laying it in a careless pile on the floor. He recognized a pair of idiotic Martian dolls. He found a tourist map of the ruined cities of Mars. He found a menu from the Red Sands Hotel.

And below all these there was a picture album. Alice at the Red Sands. Alice at the Phobos Oasis. Alice at the Darnella Ruins. He turned the pages of the album with numb fingers. Alice in a dozen Martian settings. Some of them were dated. About two years ago. They had gone together, Alice had said, but there was no evidence of Mel's presence on any such trip.

But it was equally impossible that Alice had made the trip, yet here was proof. Proof that swept him up in a doubting of his own senses. How could such a thing have taken place? Had he actually made such a trip and been stripped of the memory by some amnesia? Maybe he had forced himself to go with her and the power of his lifelong phobia had wiped it from his memory.

And what did it all have to do—if anything—with the unbelievable thing Dr. Winters had found about Alice?

Overcome with grief and exhaustion he sat fingering the mementos aimlessly while he stared at the pictures and the ticket envelope and the souvenirs.

DR. WINTERS spoke a little more sharply than he intended. "I don't think anything is going to be solved by a wild-goose chase to Mars. It's going to cost you a great deal of money, and there isn't a single positive lead to any solution."

"It's the only possible explanation." Mel persisted. "Something happened on Mars to change her from what she once was to—what you saw on your operating table."

"And you are hoping that in some desperate way you will find there was a switch of personalities—that there may be a ghost of a chance of finding Alice still alive."

Mel bit his lip. He was scarcely willing to admit such a hope but it was the foundation of his decision. "I've got to do what I can," he said. "I must take the chance. The uncertainty will torment me all my life if I don't."

Dr. Winters shook his head. "I still wish I could persuade you against it. You will find only disappointment."

"My mind is made up. Will you help me or not?"

"What can I do?"

"I can't go into space unless I can find some way of lifting, even temporarily, this phobia that nearly drives me crazy at the thought of going out there. Isn't there a drug, a hypnotic method, or something to help a thing like this?"

"This isn't my field," said Dr. Winters. "But I suspect that the cause of your trouble cannot be suppressed. It will have to be lifted. Psycho-recovery is the only way to accomplish that. I can recommend a number of good men. This, too, is very expensive."

"I should have done it for Alice—long ago," said Mel.

DR. MARTIN, the psychiatrist, was deeply interested in Mel's problem. "It sounds as if it is based on some early trauma, which has long since been wiped from your conscious memory. Recovery may be easy or difficult, depending on how much suppression of the original event has taken place."

"I don't even care what the original event was," said Mel, "if you rid me of this overwhelming fear of space. Dr. Winters said he thought recovery would be required."

"He is right. No matter how much overlay you pile on top of such a phobia to suppress it, it will continue to haunt you. We can make a trial run to analyze the situation, and then we can better predict the chance of ultimate success."

As a reporter, Mel Hastings had had vague encounters with the subject of psycho-recovery, but he knew little of the details about it. He knew it involved some kind of a machine that could tap the very depths of the human mind and drag out the hidden debris accumulated in mental basements and attics. But such things had always given him the willies. He steered clear of them.

When Dr. Martin first introduced him into the psycho-recovery room his resolution almost vanished. It looked more like a complex electronic laboratory than anything else. A half dozen operators and assistants in nurses' uniforms stood by.

"If you will recline here—," Dr. Martin was saying.

Mel felt as if he were being prepared for some inhuman biological experiment. A cage of terminals was fitted to his head and a thousand small electrodes adjusted to contact with his skull. The faint hum of equipment supported the small surge of apprehension within him.

After half an hour the preparations were complete. The level of lights in the room was lowered. He could sense the operators at their panels and see dimly the figure of Dr. Martin seated near him.

"Try to recall as vividly as possible your last experience with this nightmare you have described. We will try to lock on to that and follow it on down."

This was the last thing in the world Mel wanted to do. He lay in agonized indecision, remembering that he had dreamed only a short time ago, but fighting off the actual recollection of the dream.

"Let yourself go," Dr. Martin said kindly. "Don't fight it—"

A fragment of his mind let down its guard for a brief instant. It was like touching the surface of a whirlpool. He was sucked into the sweeping depths of the dream. He sensed that he cried out in terror as he plunged. But there was no one to hear. He was alone in space.

Fear wrapped him like black, clammy fur. He felt the utter futility of even being afraid. He would simply remain as he was, and soon he would cease to be.

But they were coming again. He sensed, rather than saw them. The searchers. And his fear of them was greater than his fear of space alone. He moved. Somehow he moved, driving headlong through great vastness while the pinpoints of light grew behind him.

"Very satisfactory," Dr. Martin was saying. "An extremely satisfactory probe."

His voice came through to Mel as from beyond vast barriers of time and space. Mel felt the thick sweat that covered his body. Weakness throbbed in his muscles.

"It gives us a very solid anchor point," Dr. Martin said. "From here I think we run back to the beginning of the experience and unearth the whole thing. Are you ready, Mr. Hastings?"

Mel felt too weak to nod. "Let 'er rip!" he muttered weakly.

THE day was warm and sunny. He and Alice had arrived early at the spaceport to enjoy the holiday excitement preceding the takeoff. It was something they had both dreamed of since they were kids—a vacation in the fabulous domed cities and ruins of Mars.

Alice was awed by her first close view of the magnificent ship lying in its water berth that opened to Lake Michigan. "It's *huge*—how can such an enormous ship ever get off the Earth?"

Mel laughed. "Let's not worry about that. We know it does. That's all that matters." But he could not help being impressed, too, by the enormous size and the graceful lines of the luxury ship. Unlike Alice, he was not seeing it at close range for the first time. He had met the ship scores of times in his reporting job, interviewing famous and well-known personages as they departed or arrived from the fabulous playgrounds of Mars.

"If you look carefully," Mel pointed out, "you'll see a lot of faces that make news when they come and go."

Alice's face glowed as she clung to Mel's arm and recognized some of the famous citizens who would be their fellow passengers. "This is going to be the most fun we've ever had in our lives, darling."

"Like a barrel of monkeys," Mel said casually, enjoying the bubbling excitement that was in Alice.

The ship was so completely stabilized that the passengers did not even have to sit down during takeoff. They crowded the ports to watch the land and the water shoot past as the ship skimmed half the length of Lake

Michigan in its takeoff run. As it bore into the upper atmosphere on an ever-increasing angle of climb, its own artificial gravity system took over and gave the illusion of horizontal flight with the Earth receding slowly behind.

Mel and Alice wandered through the salons and along the spacious decks as if in some fairyland-come-true. All sense of time seemed to vanish and they floated with the great ship in timeless, endless space.

He wasn't quite certain when he first became aware of his own sense of disquietude. It seemed to result from a change in the members of the crew. On the morning of the third day they ceased their universal and uninterrupted concern for their passengers' entertainment and enjoyment.

Most of the passengers seemed to have taken no note of it. Mel commented to Alice. She laughed at him. "What do you expect? They've spent two full days showing us the ship and teaching us to play all the games aboard. You don't expect them to play nurse to us during the whole trip, do you?"

It sounded reasonable. "I suppose so," said Mel dubiously. "But just what *are* they doing? They all seem to be in such a hurry to get somewhere this morning."

"Well, they must have some duties to perform in connection with running the ship."

Mel shook his head in doubt.

ALICE joined him in wandering about the decks, kibitzing on the games of the other passengers, and watching the stars and galaxies on the telescopic screens. It was on one of these that they first saw the shadow out in space. Small at first, the black shadow crossed a single star and made it wink. That was what caught Mel's attention, a winking star in the dead night of space.

When he was sure, he called Alice's attention to it. "There's something moving out there." By now it had shape, like a tiny black bullet.

"Where? I don't see anything."

"It's crossing that patch of stars. Watch, and you can see it blot them out as it moves."

"It's another ship!" Alice exclaimed. "That's exciting! To think we're passing another ship in all this great emptiness of space! I wonder where it's coming from?"

"And where it's going to."

They watched its slow, precise movement across the stars. After several minutes a steward passed by. Mel hailed him and pointed to the screen. "Can you tell us what that other ship is?"

The steward glanced and seemed to recognize it instantly. But he paused in replying. "That's the Mars liner," he said finally. "In just a few minutes the public address system will announce contact and change of ship."

"Change of ship?" Mel asked, puzzled. "I never heard anything about a change of ship."

"Oh, yes," the steward said. "This is only the shuttle that we're on now. We transfer to the liner for the remainder of the trip. I'm sure that was explained to you at the time you purchased your tickets." He hurried away.

Mel was quite sure no such thing had been explained to him when he purchased tickets. He turned back to the screen and watched the black ship growing swiftly larger now as it and the Martian Princess approached on contact courses.

The public address system came alive suddenly. "This is your Captain. All passengers will now prepare to leave the shuttle and board the Mars liner. Hand luggage should be made ready. All luggage stowed in the hold will be transferred without your attention. It has been a pleasure to have you aboard. Contact with the liner will be made in fifteen minutes."

From the buzz around him Mel knew that this was as much a surprise to everyone else as it was to him, but it was greeted with excitement and without question.

Even Alice was growing excited now and others crowded around them when it was discovered what they were viewing. "It looks *big*," said Alice in subdued voice. "Bigger than this ship by far."

Mel moved away and let the others have his place before the screen. His sense of uneasiness increased as he contemplated the approach of that huge black ship. And he was convinced its color *was* black, that it was not just the monotone of the view screen that made it so.

Why should there be such a transfer of passengers in mid-space? The Martian Princess was certainly adequate to make the journey to Mars. Actually they were more than a third of the way there, already. He wasn't sure why he felt so certain something was amiss. Surely there was no possibility that the great Connemorra Lines would plan any procedure to the detriment of the more than five thousand passengers aboard the ship. His uneasiness was pretty stupid, he thought.

But it wouldn't go away.

He returned to the crowd clustered at the viewing screen and took Alice by the arm to draw her away.

She looked quizzically at him. "This is the most exciting thing yet. I want to watch it."

"We haven't got much time," Mel said. "We've got a lot of things to get in our suitcases. Let's go down to our stateroom."

"Everyone else has to pack, too. There's no hurry."

"Fifteen minutes, the Captain said. We don't want to be the last ones."

Unwillingly, Alice followed. Their stateroom was a long way from the salon. The fifteen minutes were almost up by the time they reached it.

MEL closed the door to their room and put his hands on Alice's shoulders. He glanced about warily. "Alice — I don't want to go aboard that ship. There's something wrong about this whole thing. I don't know what it is, but we're not going aboard."

Alice stared at him. "Have you lost your mind? After all our hopes and all our planning you don't want to go on to Mars?"

Mel felt as if a wall had suddenly sprung up between them. He clutched Alice's shoulders desperately in his hands. "Alice — I don't think that ship out there is going to Mars. I know it sounds crazy, but please listen to me — we weren't told anything about the Martian Princess being merely a shuttle and that we'd transfer to another ship out here. No one was told. The Martian Princess is a space liner perfectly capable of going to Mars. There's no reason why such a huge ship should be used merely as a shuttle."

"That ship out there is bigger."

"Why? Do we need any more room to finish the journey?"

Alice shook herself out of his grasp. "I don't know the answers to those questions and I don't care to know them!" she said angrily.

"If you think I'm going to give up this vacation and turn right around here in space and go back home you're crazy. If you go back you'll go back alone!"

Alice whirled and ran to the door. Mel ran after her, but she was through the door and was melting into the moving throng by the time he reached the door. He took a step to follow, then halted. He couldn't drag her forcibly back into the stateroom. Maybe she'd return in a few minutes to pack her bags. He went back in the room and closed the door.

EVEN as he did so he knew that he was guessing wrong. Alice would be matching him in a game of nerves. She'd go on to the other ship, expecting him to pack the bags and follow. He sat down on the bed and put his head in his hands for a moment. A faint shudder passed through the ship and he heard the hollow ring of clashing metal. The unknown ship had made contact with the Martian Princess. Their airlocks were being mated now.

From the porthole he could see the incredible mass of the ship. He crossed the room and pressed the curtains aside. His impression had been right. The ship *was* black. Black, nameless, and blind. No insignia or portholes were visible anywhere on the hull within his range of vision.

He didn't know what he was going to do, but he knew above all else that he wasn't going to board that ship. He paced the floor telling himself it was a stupid, neurotic apprehension that filled his mind, that the great Connemorra Lines could not be involved in any nefarious acts involving five thousand people — or even one person. They couldn't afford such risk.

He couldn't shake it. He was certain that, no matter what the cost, he was not going to board that black ship.

He looked about the stateroom. He couldn't remain here. They'd certainly find him. He had to hide somewhere. He stood motionless, staring out the porthole. There was no place he could hide with assurance inside the ship.

But what about outside?

His thoughts crumpled in indecision as he thought of Alice. Yet whatever the black ship meant, he could help no one if he went aboard. He had to get back to Earth and try to find out what it was all about and alert the authorities. Only in that way could he hope to help Alice.

HE opened the stateroom door cautiously and stepped out. The corridor was filled with hurrying passengers, carrying hand luggage, laughing with each other in excitement. He joined them, moving slowly, alert for crew members. There seemed to be none of the latter in the corridors.

Keeping close to the wall, he moved with the crowd until he reached the rounded niche that marked an escape chamber. As if pushed by the hurrying throng, he backed into it, the automatic doors opening and closing to receive him.

The chamber was one of scores stationed throughout the ship as required by law. The escape chambers contained space suits for personal exit from the ship in case of emergency. They were never expected to be used. In any emergency requiring abandonment of the vessel it would be as suicidal to go into space in a suit as to remain with the ship. But fusty lawmakers had decreed their necessity, and passengers received a perfunctory briefing in the use of the chambers and the suits — which they promptly forgot.

Mel wrestled now with what he remembered of the instructions. He inspected a suit hanging in its cabinet and then was relieved to find that the instructions were repeated on a panel of the cabinet. Slowly, he donned the suit, following the step by step instructions as he went. He began to sweat profusely from his exertions and from his fear of discovery.

He finally succeeded in getting the cumbersome gear adjusted and fastened without being detected. He did not know if the airlock of the chamber had some kind of alarm that would alert the crew when it was opened. That was a chance he had to take. He discovered that it was arranged so that it could be opened only by a key operated from within the suit. This was obviously to prevent anyone leaving the ship unprotected. Perhaps with this safeguard there was no alarm.

He twisted the lock and entered the chamber. He opened the outer door and faced the night of space.

He would not have believed that anything could be so utterly terrifying. His knees buckled momentarily and left him clinging to the side of the port. Sweat burst anew from every pore. Blindly, he pressed the jet control and forced himself into space.

He arced a short distance along the curve of the ship and then forced himself down into contact with the hull. He clung by foot and hand magnetic pads, sick with nausea and vertigo.

He had believed that by clinging to the outside of the hull he could escape detection and endure the flight back to Earth. In his sickness of body and mind the whole plan now looked like utter folly. He retched and closed his eyes and lay on the hull through the beginning of an eternity.

HE had no concept of time. The chronometer in the suit was not working. But it seemed as if many hours had passed when he felt a faint shock pass through the hull beneath him. He felt a momentary elation. The ships had separated. The search for him — if any — had been abandoned.

Slowly he inched his way around the hull to get a glimpse of the black ship. It was still there, standing off a few hundred yards but not moving. Its presence dismayed him. There could be no reason now for the two ships to remain together. The Martian Princess should be turning around for the return to Earth.

Then out of the corner of his eye he saw it. A trace of movement. A gleam of light. Like a small moon it edged up the distant curvature of the hull. Then there were more — a nest of quivering satellites.

Without thought, Mel pressed the jet control and hurled himself into space.

The terror of his first plunge was multiplied by the presence of the searchers. Crewmen of the Martian Princess, he supposed. The absence of the space suit had probably been discovered.

In headlong flight, he became aware of eternity and darkness and loneliness. The sun was a hot, bright disc, but it illuminated nothing. All that his mind clung to for identification of itself and the universe around it was gone. He was like a primeval cell, floating without origin, without purpose, without destination.

Only a glimmer of memory pierced the thick terror with a shaft of rationality. Alice. He must survive for Alice's sake. He must find the way back to Alice — back to Earth.

He looked toward the Martian Princess and the searchers on the hull. He cried out in the soundless dark. The searchers had left the hull and were pursuing him through open space. Their speed far exceeded his. It was futile to run before them — and futile to leave the haven of the Martian Princess. His only chance of survival or success lay in getting to Earth aboard the ship. In a long curve he arced back toward the ship. Instantly, the searchers moved to close in the arc and meet him on a collision course.

He could see them now. They were not crewmen in spacesuits as he had supposed. Rather, the objects — two of them — looked like miniature spaceships. Beams of light bore through space ahead of them, and he suspected they carried other radiations also to detect by radar and infra red.

In the depths of his mind he knew they were not of the Martian Princess. Nor were there any crewmen within them. They were robot craft of some kind, and they had come from the great black ship.

He felt their searching beams upon him and waited for a deadly, blasting burst of heat or killing radiation. He was not prepared for what happened.

THEY closed swiftly, and the nearest robot came within a dozen feet, matching Mel's own velocity. Suddenly, from a small opening in the machine, a slender metal tentacle whipped out and wrapped about him like the coils of a snake. The second robot approached and added another binding. Mel's arms and legs were pinned. Frantically, he manipulated the jet control in the glove of the suit. This only caused the tentacles to cut deeply and painfully, and threatened to smash the shell of the suit. He cut the jets and admitted the failure of his frantic mission.

In short minutes they were near the ships again. Mel wondered what kind of reprimand the crewmen of the Martian Princess could give him, and what fantastic justification they might offer for their own actions.

But he wasn't being taken toward the Martian Princess. He twisted painfully in the grip of the robot tentacles and confirmed that he was being carried to the black stranger.

Soundlessly, a port slid open and the robots swept him through into the dark interior of the ship. He felt himself dropped on a hard metal floor. The tentacles unwound. Alone, he struggled to his feet and flashed a beam of light from the suit flashlight to the walls about him. Walls, floor and ceiling were an indistinguishable dark gray. He was the only object in the chamber.

While he strained his sight to establish features in the blank metallic surfaces a clipped, foreign voice spoke. "Remove your suit and walk toward the opening in the wall. Do not try to run or attack. You will not be harmed unless you attack."

There was no use refusing. He did as commanded. A bright doorway opened in the wall before him. He walked through.

It reminded him of a medical laboratory. Shelves and cabinets of hand instruments and electronic equipment were about. And in the room three men sat watching the doorway through which he entered. He gazed at the strangers as they at him.

They looked ordinary enough in their white surgeons' smocks. All seemed to be of middle age, with dark hair turning gray at the fringe. One was considerably more muscular than the other two. One leaned to overweight. The third was quite thin. Yet Mel felt himself bristling like a dog in the dark of the moon.

No matter how ordinary they looked, these three were not men of Earth. The certainty of this settled like a cold, dead weight in the pit of his stomach.

"You—" he stammered. There was nothing to say.

"Please recline on this couch," the nearest, the muscular one said. "We wish you no harm so do not be afraid. We wish only to determine if you have been harmed by your flight into space."

All three of them were tense and Mel was sure they were worried—by his escapade. Had he nearly let some unknown cat out of the bag?

"Please—," the muscular one said.

He had no alternative. He might struggle, and destroy a good deal of apparatus, but he could not hope to overwhelm them. He lay on the couch as directed. Almost instantly the overweight one was behind him, seizing his arm. He felt the sting of a needle. The thin one was at his feet, looking down at him soberly. "He will rest," the thin one said, "and then we shall know what needs to be done."

The sleep had lasted for an eon, he thought. He had a sense of the passage of an enormous span of time when he at last awoke. His vision was fuzzy, but there was no mistaking the image before him.

Alice. His Alice—safe.

She was sitting on the edge of the bed, smiling down at him. He fought his way up to a half-sitting position. "Alice!" He wept.

Afterwards, he said, "Where are we? What happened? I remember so many crazy things—the vacation to Mars."

"Don't try to remember it all, darling," she said. "You were sick. Some kind of hysteria and amnesia hit you while we were there. We're home now. You'll soon be out of the hospital and everything will be all right."

"I spoiled it," he murmured. "I spoiled it all for you."

"No. I knew you were going to be all right. I even had a lot of fun all by myself. But we're going back. As soon as you are all well again we'll start saving up and go again."

He nodded drowsily. "Sure. We'll go to Mars again and have a real vacation."

Alice faded away. All of it faded away.

AS if from a far distance the walls of Dr. Martin's laboratory seemed to close about him and the lights slowly increased. Dr. Martin was seated beside him, his head shaking slowly. "I'm so terribly sorry, Mr. Hastings. I thought we were going to get the full and true event this time. But it often happens, as in your case, that fantasy lies upon fantasy, and it is necessary to dig through great layers of them before uncovering the truth. I think, however, that we shall not have to go much deeper to find the underlying truth for you."

Mel lay on the couch, continuing to stare at the ceiling. "Then there was no great, black ship out of space?"

"Of course not! That is one danger of these analyses, Mr. Hastings. You must not be deceived into believing that a newly discovered fantasy is the truth for which you were looking. You must come back and continue your search."

"Yes. Yes, of course." He got up slowly and was helped to the outer room by the Doctor and an attendant. The attendant gave him a glass of white, sweetish substance to drink.

"A booster-upper," laughed Dr. Martin. "It takes away the grogginess that sometimes attends such a deep sweep. We will look for you day after tomorrow."

Mel nodded and stepped out into the hall.

No great black ship.

No mysterious little robot ships with tentacles that whip out and capture a man.

No strange trio in surgeons' gowns.

And no Alice—

A sudden spear of thought pierced his mind. Maybe all that was illusion, too. Maybe he could go home right now and find her waiting for him. Maybe—

No. That was real enough. The accident. Dr. Winters. The scene in the icy room next to Surgery at the hospital. Dr. Martin didn't know about that. He would have called that a fantasy, too, if Mel had tried to tell him.

No. It was all real.

The unbelievable, alien organs of Alice.

The great, black ship.

The mindless robot searchers.

His nightmare had stemmed from all this that had happened out in space, which had somehow been wiped from his conscious memory. The nightmare had not existed in his boyhood, as he had thought. It was oriented in time now.

But what had happened to Alice? There was no clue in the memory unearthed by Dr. Martin. Was her condition merely the result of some freak heredity or gene mutation?

The surging turmoil in his mind was greater than before. There was only one way to quiet it—that was to carry out his original plan to go to Mars.

He'd go out there again. He'd find out if the black ship existed or not.

THE girl in the ticket office was kind but firm. "Our records show that you were a vacationer to Mars very recently. The demand is so great and the ship capacity so small that we must limit vacation trips to no more than one in any ten-year period."

He turned away and went down the hall and out the doorway of the marble and brass Connemorra Lines Building.

He walked through town for six blocks and the thought of old Jake Norton came to his mind. Jake had been an old timer in the city room when

Mel was a cub. Jake had retired just a few months ago and lived in a place in town with a lot of other old men. Mel hailed the nearest cab and drove to Jake's place.

"Mel, it's great to see you!" Jake said. "I didn't think any of the boys would remember an old man after he'd walked out for the last time."

"People remember real easy when they want favors."

"Sure," Jake said with a grin, "but there's not much of a favor I can do you any more, boy. Can't even loan you a ten until next payday."

"Jake, you can help me," said Mel. "You don't expect to ever take a trip to Mars, do you?"

"Mars! Are you crazy, Mel?"

"I went once. I've got to go again. It's about Alice. And they won't let me. I didn't know you could go only once in ten years."

Jake remembered. Alice had called him and all the other boys after they'd come back the other time. Mel had been sick, she said. He wouldn't remember the trip. They were asked not to say anything about it. Now Mel was remembering and wanted to go again. Jake didn't know what he should do.

"What can I do to help you?" he asked.

"I'll give you the money. Buy a ticket in your name. I'll go as Jake Norton. I think I can get away with it. I don't think they make any closer check than that."

"Sure—if it'll do you any good," Jake said hesitantly. He was remembering the anxiety in Alice's voice the day she called and begged him not to say anything that would remind Mel of Mars. No one ever had, as far as Jake knew.

He took the money and Mel waited at the old men's home. An hour later Jake called. "Eight months is the closest reservation I can get at normal rates, but I know of some scalpers who charge 50% more."

Mel groaned. "Buy it no matter what the cost! I've got to go at once!" He would be broke for the next ten years.

IT was little different from the other time. There was the same holiday excitement in the crowd of vacationers and those who had come to see them off. It was the same ship, even.

All that was different was the absence of Alice.

He stayed in his stateroom and didn't watch the takeoff. He felt the faint rocking motion as the ship went down its long waterway. He felt the shift as the artificial gravity took over. He lay on the bed and closed his eyes as the Martian Princess sought the cold night of space.

For two days he remained in the room, emerging only for meals. The trip itself held no interest for him. He waited only for the announcement that the black ship had come.

But by the end of the second day it had not come. Mel spent a sleepless night staring out at the endless horizon of stars. Dr. Martin had been right, he thought. There was no black ship. He had merely substituted one illusion for another. Where was reality? Did it exist anywhere in all the world?

Yet, even if there were no black ship, his goal was still Mars.

The third day passed without the appearance of the black ship. But on the very evening of that day the speaker announced: "All passengers will prepare for transfer from the shuttle ship to the Mars liner. Bring hand luggage—"

Mel sat paralyzed while he listened to the announcement. So it was true! He felt the faint jar that rocked the Martian Princess as the two ships coupled. From his stateroom port Mel could see the stranger, black, ugly, and somehow deadly. He wished he could show Dr. Martin this "illusion"!

He packed swiftly and left the room. Mel joined the surprised and excited throng now, not hanging back, but eager to find out the secret of the great black ship.

The transition from one ship to the other was almost imperceptible. The structure of both corridors was the same, but Mel knew when the junction was crossed. He sensed the entry into a strange world that was far different from the common one he knew.

Far down the corridor the crowd was slowing, forming into lines before stewards who were checking tickets. The passengers were shunted into branching corridors leading to their own staterooms. So far everything was so utterly normal that Mel felt an overwhelming despondency. It was just as they had been told; they were transferring to the Mars liner from the shuttle.

The steward glanced at his ticket, held it for a moment of hesitation while he scanned Mel's face. "Mr. Norton—please come with me."

The steward moved away in a direction no other passengers were taking. Another steward moved up to his place. "That way," the second man said to Mel. "Follow the steward."

MEL'S heart picked up its beat as he stepped out of the line and moved slowly down the corridor after the retreating steward. They walked a long way through branching silent corridors that showed no sign of life.

They stopped at last before a door that was like a score of others they had passed. There were no markings. The steward opened the door and stood aside. "In here please," he said. Mel entered and found himself alone. The steward remained outside.

The room was furnished as an office. It was carpeted and paneled luxuriously. A door leading from a room at his left opened and admitted a tall man with graying hair. The man seemed to carry an aura of power and strength as he moved. An aura that Mel Hastings recognized.

"James Connemorra!" Mel exclaimed.

The man bowed his head slightly in acknowledgement. "Yes, Mr. Hastings," he said.

Mel was dismayed. "How do you know who I am?" he said.

James Connemorra looked through the port beside Mel and at the stars beyond. "I have been looking for you long enough I ought to know who you are."

Something in the man's voice chilled Mel. "I have been easy enough to find. I'm only a news reporter. Why have you been looking for me?"

Connemorra sank into a deep chair on the opposite side of the room. "Can't you guess?" he said.

"It has something to do with what happened—before?" Mel asked. He backed warily against the opposite wall from Connemorra. "That time when I escaped from the Martian Princess rather than come aboard the black ship?"

Connemorra nodded. "Yes."

"I still don't understand. Why?"

"It's an old story." Connemorra shrugged faintly. "A man learns too much about things he should know nothing of."

"I have a right to know what happened to my wife. You know about her don't you?"

Connemorra nodded.

"What happened to her? Why was she different after her trip to Mars?"

James Connemorra was silent for so long that Mel thought he had not heard him. "Is everyone different when they get back?" Mel demanded. "Does something happen to everybody who takes the Mars trip, the same thing that happened to Alice?"

"You learned so much," said Connemorra, speaking as if to himself, "I had to hunt you down and bring you here."

"What do you mean by that? I came through my own efforts. Your office tried to stop me."

"Yet I knew who you were and that you were here. I must have had something to do with it, don't you think?"

"What?"

"I forced you to come by deception, so that no one knows you are here — except the old man whose name you used. Who will believe him that you came on the Martian Princess? Our records will show that a Jake Norton will be there on Earth. No one can ever prove that Mel Hastings ever came aboard."

MEL let his breath out slowly. His fear suddenly swallowed caution. He took a crouching step forward. Then he stopped, frozen. James Connemorra tilted the small pistol resting in his lap. Mel did not know how it came to be there. He had not seen it a moment ago.

"What are you going to do?" Mel demanded. "What are you going to do with all of us?"

"You know too much," said Connemorra, shrugging in mock helplessness. "What *can* I do with you?"

"Explain what I don't understand about the things you say I know."

"Explain to you?" The idea seemed to amuse Connemorra greatly, as if it had some utterly ridiculous aspect. "Yes, I might as well explain," he said. "I haven't had anyone interested enough to listen for a long time.

"Men have never been alone in space. We have been watched, inspected, and studied periodically since Neanderthal times by races in the galaxy who have preceded us in development by hundreds of thousands of years. These observers have been pleasantly excited by some of the things we have done, appalled by others.

"There is a galactic organization that has existed for at least a hundred thousand years. This organization exists for the purpose of mutual development of the worlds and races of the galaxy. It also exists to maintain peace, for there were ages before its organization when interstellar war took place, and more than one great world was wiped out in such senseless wars.

"When men of Earth were ready to step into space, the Galactic Council had to decide, as it had decided on so many other occasions, whether the new world was to be admitted as a member. The choice is not one which a new world is invited to make; the choice is made for it. A world which begins to send its ships through space becomes a member of the Council, or its ships cease to travel. The world itself may cease to exist."

"You mean this dictatorial Council determines whether a world is fit to survive and actually wipes out those it decides against?" gasped Mel in horror. "They set themselves up as judges in the Universe?"

"That's about the way they operate, to put it bluntly," said Connemorra. "You can call them a thousand unpleasant names, but you can't change the fact of their existence, nor the fact of their successful operation for a period as long as the age of the human race.

"They would never have made their existence known to us if we had not begun sending our ships into space. But once we did that we were entering territory staked out by races that were there when we crawled out of our caves. Who can say what their rights are?"

"But to pass judgment on entire worlds—"

"We have no choice but to accept that such judgment is passed."

"And their judgment of Earth—?"

"Was that Earth was not ready for Council membership. Earthmen are still making too many blunders to join creatures that could cross the galaxy at the speed of light when we were learning how to chip flint."

"But they didn't wipe us out!"

JAMES CONNEMORRA looked out at the stars. "I wonder," he said. "I wonder—"

"What do you mean?" Mel said in a tight voice.

"We have defects which are not quite like any they have encountered before. We have developed skills in the manufacture of artifacts, but we have no capacity for using them. For example, we have developed vast systems of communication, but these systems have not improved our communications they have actually blocked communication."

"That's crazy!" said Mel. "Do they suppose smoke signals are superior to the 3-d screens in our homes?"

"As a matter of fact, they do. And so do I. When a man must resort to smoke signals he is very certain that he has something to say before he goes to the trouble of putting the message in the air. But our fabulous screens prevent us from communicating with each other by throwing up a wall of pseudo-communication that we can't get through. We subject ourselves to a barrage of sound and light that has a communication content of almost zero.

"The same is true of our inventions in transportation. We have efficient means of travel to all parts of the world and now to the Universe itself. But we don't travel. We use our machines to block traveling."

"I can understand the first argument, but not this one!" said Mel.

"We move our bodies to new locations with our machines, but our minds remain at home. We take our rutted thoughts, our predispositions, our cultural concepts wherever we go. We do not touch, even with a fragment of our minds, that which our machines give us contact with. We do not travel. We move in space, but we do not travel.

"This is their accusation. And they're right. We are still doing what we have always done. We are using space flight for the boring, the trivial, the stupid; using genius for a toy, like a child banging an atomic watch on the floor. It happened with all our great discoveries and inventions: the gasoline engine, the telephone, the wireless. We've built civilizations of monumental stupidity on the wonders of nature. One race of the Galactics has a phrase they apply to people like us: 'If there is a God in Heaven He has wept for ten thousand years.'

"BUT all this is not the worst. A race that is merely stupid seldom gets out to space. But ours has something else they fear: destructiveness. They have plotted our history and extrapolated our future. If they let us come out, war and conflict will follow."

"They can't know that!"

"They say they can. We are in no position to argue."

"So they plan to destroy us—"

"No. They want to try an experiment that has been carried out just a few times previously. They are going to reduce us from what they term the critical mass which we have achieved."

"Critical mass? That's a nuclear term."

"Right. Meaning ready to blow up. That's where we are. Two not-so-minor nuclear wars in fifty years. They see us carrying our destructiveness into space, fighting each other there, infecting other races with our hostility. But if we are broken down into smaller groups, have the tools of war removed, and are forced to take another line of development—well, they have hopes of salvaging us."

"But they can't do a thing like that to us! What do they intend? Taking groups of Earthmen, deporting them to other worlds—breaking them apart from each other forever—?"

The coldness found its resting place in Mel's chest. He stared at James Connemorra. Then his eyes moved slowly over the walls of the room in the black ship and out to the stars. The black ship.

"This ship—! You transfer your passengers to this Galactic ship for deportation to other worlds! But they come back—"

"They are sent to colonies on other worlds where conditions are like those on Earth—with significant exceptions. The colonies are small, the largest are only a few thousand. The problems there are different than on Earth—and they are tough. The natural resources are not the same. The development of the resulting cultures will be vastly different from that of Earth. The Galactic Council is very interested in the outcome—which will not be known with certainty for a thousand years or so."

"But they come back," Mel repeated. "You bring them back!"

"For each Earthman who goes out, a replacement is sent back. The replacement is an android supplied by the Council."

"Android!" Mel felt his reason slipping. He knew he was shouting. "Then Alice—the Alice that died was an android, she was not my wife! My Alice is still alive! You can take me to her—"

Connemorra nodded. "Alice is still alive, and well. No harm has come to her."

"Take me to her!" Mel knew he was pleading, but in his anguish he had no pride.

Connemorra seemed to ignore his plea. "Earth's population is slowly being diluted by the removal of top people. The androids behave in every way like the individuals they replace, but they are preconditioned against the inherent destructiveness of Earthmen."

Blind anger seemed to rise within Mel. "You have no right to separate me from Alice. Take me to her!"

His rage ignited and he leaped forward.

The small gun in Connemorra's hand spurted twice. Mel felt a double impact in a moment of great wonder. It couldn't end like this, he thought. It couldn't end without his seeing Alice once more. Just once more—

HE sank to the floor. The pain was not great, but he knew he was dying. He looked down at his hand that covered the great wound in his mid-section. There was something wrong.

He felt the stickiness, but the red blood was not welling out. Instead, a thick bubble of green ooze moved from the wound and spread over his clothes and his hand. An alien greenness that was like nothing human.

He had seen it once before.

Alice.

He stared up at Connemorra with wide, wondering eyes.

"Everything went wrong, my poor android," said Connemorra softly. "After your human was brought back to the ship we were forced to go through with the usual process of imprinting his mind content upon his android. But we had to wipe out all memory of the attempted escape from the Martian Princess. This was not successful. It still clung in the nightmares you experienced. And the psycho-recovery brought it all back.

"We tried to cover it with an amnesiac condition instead of the usual pre-printed memory of a Mars vacation. And all this might have worked if the Alice android had not been defective also. A normal android has protective mechanisms that make accidents and subsequent discovery impossible. But the Alice android failed, and you set out on a course to uncover us. I had to find a way to destroy you—murder."

"I'm truly sorry. I don't know how an android thinks or feels. Sometimes I'm afraid of all of you. You are like men, but I've seen the factories in which you are produced. There are many things I do not know. I know only that I had to obey the Galactic Council or Earth would have been destroyed long ago.

"And something else I know: Alice and Mel Hastings are content and happy. They are on a lovely world, very much like Central Valley."

He closed his eyes as he felt the life — or whatever it was — seeping out of him. It came out right, after all, he thought.

Like a wooden soldier with a painted smile, fallen from a shelf, he lay twisted upon the floor.